Executive Summary

Section 1406 of the Energy Policy Act (EPACT) of 2005, directs the Secretary of Energy to conduct a study of the application of radiation to petroleum at standard temperature and pressure to refine petroleum products, whose objective shall be to increase the economic yield from each barrel of oil. The goals of the study shall include: (1) increasing the value of our current oil supply; (2) reducing the capital investment cost for cracking oil; (3) reducing the operating energy cost for cracking oil; and (4) reducing sulfur content using an environmentally responsible method. However, funding was not appropriated for the study. The Department of Energy (DOE) Office of Oil and Natural Gas decided to conduct a limited study to review and analyze available literature on cold cracking technology and its application to crude refining.

Purpose of Work

The main purpose of this limited study was to conduct a literature search on "cold cracking" of petroleum crude oil in an attempt to collect and analyze background information.

Methodology

Available literature (papers, patents, and reports) on radiation science and technology and its application to hydrocarbon cracking/processing obtained from various sources were reviewed. Private discussions with many researchers and experts in the subject area were also held. Input on the anticipated cost of radiation sources was also obtained.

Major Findings

It is obvious from the literature review and private discussions with experts that application of radiation cracking of larger molecules found in crude oils to make (refine) higher value lighter petroleum products is not new. Radiation in different forms (neutrons, electrons, X-rays, Gamma-rays, etc.) can be delivered to petroleum crude oil with energy that is many orders of magnitude in excess of that required to break large hydrocarbon molecules. Some specific observations from the literature include the following:

- Radiation research applied to petroleum has been pursued by major oil companies, academia and National Laboratories since the 1950's. A wide variety of radiation sources have been explored as high energy sources: emitting neutrons, electrons, gamma-ray, beta particles and fission products. There are a few energy company U.S. patents that provide detailed results comparing yields from radiation processing of crude oil with conventional refined product yields. None of the patents proved commercial because the energy balance was uncertain.

- Interest in radiation processing of petroleum languished for about 30 years. Most energy companies and refiners disbanded their nuclear divisions by 1983. Government and academic research on nuclear and hydrocarbons (bond breaking and molecular rearrangement) continued but was limited to model hydrocarbon compounds – clean, well-defined multi-component systems. However, discussion of the use of excess heat from nuclear reactors continued to be of interest to both energy companies and academia.

- Renewed interest in application of radiation to petroleum processing was generated after the collapse of the former Soviet Union, when many Russian nuclear scientists became idle. Since the 1990's, international efforts, through the International Atomic Energy Agency (IAEA) led to funding a series of proposals from the nuclear scientists in the

Republic of Kazakhstan. The use of high energy electrons, generated by a linear electron accelerator, resulted in a series of publications for the Kazakh scientists. Some studies indicate improved efficiency, and highlighted some of the limitations of using high flux electrons to process different types of crude oils found in central Russia and Kazakhstan.

- This recent Kazakh work has generated renewed interest from a number of U.S. companies that hope to patent the technology and further define the viability of using electron beams as a way of refining crude oil to produce higher value liquid fuels.

Much of the chemistry in the recent electron beam work on irradiating molecules is well known. Volumes of books and articles have been written on free radical chemistry, and radiolysis chemistry in well defined molecular systems has been studied extensively. However, the chemistry becomes significantly more complicated when dealing with complex mixtures such as crude oil. A huge number of radicals are formed when electrons (from a linear accelerator) collide with organic molecules. Reactions with well-defined organic molecules or families of molecules have been studied. Reactions with the broad mixture, similar to that in crude oils, have received much less attention. The key is controlling the generation and propagation of the radicals to minimize polymerization, which produces higher molecular weight materials – asphalts or coke. Controlling mechanisms are not well defined. Refined products tend to be unstable (react during storage to yield poorer quality fuels). Some recently published papers comment that fuels processed using an electron beam degraded with time. Kazakh experimental research cited in many recent papers on use of radiation (electrons) were not performed at standard temperature and pressure (STP). They were conducted at a few hundred degrees less than conventional thermal refining conditions, and required pressure in order to confine generated gases.

Data within recent publicly available reports on the use of electron beams indicate progress is being made toward defining some of the limitations in the use of radiation (electrons and gamma-ray). Recent Kazakh research focuses on batch operation, where an electron accelerator is used to expose a batch of crude oil – a laboratory "beaker or flask" sample. The publications do not clearly define operating conditions or product analysis details. For example,

- To make valid comparisons, the best current conventional thermal refining processes need to be compared with the best radiation refining process (combined with thermal processing for product separation) using well-characterized crude oils samples that are processed by U.S. refiners.
- For "cold cracking", it would be anticipated that refinery streams currently sent to a fluidized catalytic cracker (FCC) unit would be treated, and not the entire crude oil stream since irradiating short hydrocarbon molecules tends to polymerize molecules and form higher molecular weight, lower value fuels.
- Better and more detailed characterization of the yield, quality, and stability of products obtained when using high energy electrons needs to be done and compared with samples from current thermal refining processes in order to assess the viability of radiation in obtaining higher yield of equal or better quality products.
- Analysis of yield and quality of products obtained via radiation must be based on continuous flow systems rather than batch (laboratory beaker-scale) systems to make more accurate estimates of both capital and operating cost for a reliable comparison with conventional refining. The estimates for radiation processing have to include the critical processing parts of a refinery that are used for separation of molecules into various

boiling ranges, molecular rearrangement and production of commercial fuels - because they are needed independent of energy source (radiation or thermal).

- Rough cost comparisons for construction of an electron accelerator in conjunction with a refinery for product separation versus a grass-roots deep-conversion refinery (minimum size 100,000 bbl/day) would help determine capital costs.

- Operating costs of a linear accelerator are less well known than those for conventional oil refineries. The only cost component given for radiation is the cost of generating the radiation energy. Information on other operating cost components such as labor, utilities, etc is usually not available. Most accelerators are heavily government subsidized and are used for academic fundamental research rather than commercial product production.

- Processing of crude or refined petroleum products at standard temperature and pressure (STP), defined as a temperature of 0 degrees Celsius or 32 degrees Fahrenheit, and a pressure of 1 atmosphere or 14.7 pound per square inch, is uncertain. This is because of material handling problems associated with such multi-phase (gas, liquid and solid) systems. Lighter petroleum molecules (like methane, ethane propane and butane) are gases at STP. Intermediate molecules (pentane, hexane, octane, gasoline, kerosene and diesel boiling range fractions) are liquids. Large molecules (dodecane, etc.) like asphalt and wax are solids. Efficient handling of all three phases at STP in large volumes, as required in a refinery, is not feasible.

- Review results indicate that processing crude oil in a cost-effective and environmentally responsible manner using radiation would be difficult.

- Limited and inconsistent literature data indicate the economics of "cold cracking" are uncertain.

- Technology has not yet been adequately tested. The few cold cracking "beaker tests" were not comprehensively studied. There is too little quantitative information to make comparisons with conventional refining. It is therefore unclear if there are energy benefits if cold cracking were used.

- Refining industry's acceptance of the technology is uncertain. New technology is difficult to introduce into the refining industry.

- The radiation energy from a given source varies widely. The energies available from radiation are orders of magnitude higher than the activation energy needed for breaking a carbon-carbon, carbon-hydrogen, carbon-sulfur, or carbon-nitrogen bond. However, with conventional thermal crude oil processing, just enough energy to overcome the energy of activation is supplied to break the bond or conduct the chemical reaction.

Potential of Meeting Section 1406 of EPACT Study Goals

1. **Increasing the value of our current oil supply:** It is not obvious that application of radiation to petroleum processing can increase the value of our current oil supply. This is because electrical efficiency in linear accelerators is poor. It is difficult to determine the efficiency of crude oil radiation processing to make refined products with the available data. Improved product yield, product quality, product stability and/or lower energy use have not simultaneously been demonstrated in the literature.

2. **Reducing the capital investment cost for cracking oil:** There is no obvious evidence in the literature to determine the ability of radiation to reduce the capital cost for refining crude oil. The cost of a linear accelerator and the associated thermal distillation units to separate the separate fractions that are produced would probably exceed the cost of a deep-conversion conventional refinery, even for a new grass-roots refinery. Permitting a

dedicated radiation source in or near a refinery is uncertain and may take a long time. Delays would add to the cost of construction.

3. **Reducing the operating energy cost for cracking oil:** Recent literature has not shown sufficient details and/or assumptions made on estimating the total energy cost required to process a given volume of crude oil. At best, claims in recent papers are specious. For example, some of the estimates for "processing" are less than what is required to heat heavy crude to make it pumpable to the first reaction vessel.

4. **Reducing sulfur content using an environmentally responsible method:** Use of radiation has been shown to break larger organic sulfur molecules into low molecular weight, low sulfur-containing products (gasoline and diesel fraction), while the sulfur content of higher molecular weight products (asphalt, vacuum bottoms) increases. Sulfur atoms do not get converted to other atoms in a high flux electron beam - "cold cracking". Conventionally, sulfur in the crude oil is usually chemically converted to some intermediate compounds, such as H_2S, and finally to elemental sulfur.

Energy Policy Act of 2005 Section 1406

This report constitutes the collection and analysis of background information from the literature on "Cold Cracking of Petroleum Crude Oil" as described in Section 1406 of the Energy Policy Act (EPACT) of 2005. Section 1406 of EPACT 2005 states that: *"The Secretary (of Energy) shall conduct a study of the application of radiation to petroleum at standard temperature and pressure to refine petroleum products, whose objective shall be to increase the economic yield from each barrel of oil. The goals of the study are (1) increasing the value of our current oil supply; (2) reducing the capital investment cost for cracking oil; (3) reducing the operating energy cost for cracking oil; and (4) reducing sulfur content using an environmentally responsible method"*.

Methodology

Available literature (papers, patents, and reports) on radiation science and technology and its application to hydrocarbon cracking/processing obtained from various sources were reviewed. Private discussions with many researchers and experts in the subject area were also held.

Review of the literature and discussion with radiation experts at the DOE's National Laboratories and U.S. universities was pursued as part of this study. The literature review was not intended to be exhaustive, but rather a selection of pertinent articles. The intent of this review is to provide background information on the application of radiation to petroleum refining and define needed subsequent steps for this study. Since engineers and operators of conventional refineries are usually unfamiliar with radiation chemistry and since radiation physicists and chemists may be equally unfamiliar with conventional refining practices, literature on the basics of both areas has been reviewed or incorporated by reference.

As the literature review progressed, the focus shifted from all radiation sources to application of high energy electrons - because of advancement in electron beams in recent decades. One can now nearly dial in the energy of the electron and chose an energy that maximizes gain in certain reactions. Cold cracking of petroleum by use of sonication, microwaves or microwave heating (which has numerous citations) was omitted from this literature search.

Both Petroleum Abstracts and Chemical Abstracts (CAPLUS) were searched for the terms "radiation and refining" using the University of Tulsa library. Significant published work existed in the Petroleum and Chemical Abstracts in Russian and other foreign languages. These publications were not pursued, and only the ones in English were reviewed.

The U.S. patent literature was searched by keywords using the online search capabilities at the U.S. Patent and Trademark Office. Keyword search of U.S. patents from 1975 to present, available at www.uspto.gov, was searched using the following keywords "cold cracking, electron radiation induced chemical conversion, radiation chemical conversion, radiation thermal cracking, and radiation and refining." Selected patents were reviewed, and related patents cited in the original documents were also reviewed. Search of U.S. patents from 1950 to 1975 for the terms "radiation and oil" was conducted resulting in over 3000 citations, only a few of which were applicable.

A previous literature search by Nerac (2005) was limited by restricting temperatures to less than $300^\circ C$, and to atmospheric pressure. This eliminates most of the patents and recent papers on radiation-thermal conversion using high energy electrons produced by an electron accelerator.

The most productive literature search using the above terms was via the Journal of Radiation Physics and Chemistry, accessible at www.sciencedirect.com. This Journal has become the major publishing site for much of the work on use of radiation physics, radiation biology and radiation chemistry in the last two decades.

Search of the International Atomic Energy Agency (IAEA) website http://inis-a4.iaea.org/demo/php/ was conducted for "oil and radiation." It yielded a number of proposals and progress reports on irradiation of crude oil, petroleum fractions and/or model compounds.

A search of the International Science and Technology Center (ISTC, Russia) databases for "oil and refining, or oil and radiation" at http://search.istc.ru yielded nearly the same proposals and summaries of reports. ISTC is partially funded by International Atomic Energy Agency.

Discussions with U.S researchers working on high radiation sources and radiochemistry suggest that there are classified studies within DOE's National Laboratories (not accessible and not reviewed).

In addition to analyzing the written literature, interviews with refining researchers and radiation researchers were conducted. The refiners represent both U.S. major refiners and international companies who have major interest in processing the "bottom of the barrel" high molecular weight oil to make higher value liquid transportation fuels. The nuclear-radiation researchers were from U.S. DOE National Laboratories with linear accelerators or research nuclear reactors and from U.S. universities that have nuclear science programs (research and teaching) and expertise in operating linear accelerators.

Background

Properties of Crude Oils

Crude oils are colloidal mixtures of a huge number of hydrocarbons. They range from condensates that are light, highly volatile, and composed of predominately low molecular weight hydrocarbons to heavy, dense, highly viscous crude (heavy oil, bitumen) with a preponderance of high molecular weight molecules. Heavy crudes usually contain more sulfur, nitrogen and metals and sell at significant discounts to light, sweet crude oils like West Texas Intermediate (WTI).

Crude oils assays are available for a variety of crude oils. Refiners use assays to estimate yield (of refined products) for their specific refinery configuration. A simple refinery can only handle light sweet (low sulfur) crude. It would encounter significant processing problems if it processed heavy sour (high sulfur) crude. This limits their selection of crude feed that this refinery can process to light sweet crudes that normally command a higher price. A deep conversion, complex refinery is designed to process heavy sour crude. It can process light sweet crudes, but may not achieve as much economic return from processing the more expensive light sweet crude.

The solubility of various molecules (or classes of molecules) and the molecular weights of molecules are a delicate balance in most crude oils. Disturbing the balance in given crude by altering the temperature, pressure, or composition causes the colloidal crude oil to lose stability and possibly, change from a single phase to multiple phases (gas, liquid and solid).

Processing of crude (or refined petroleum products) at standard temperature and pressure (STP) is uncertain. STP is defined as a temperature of 0 degrees Celsius (32 degrees Fahrenheit) and pressure of 1 atmosphere (14.7 pounds per square inch). Lighter petroleum molecules (like methane, ethane propane and butane) are gases at STP. Intermediate components (pentane, hexane, octane, gasoline, kerosene and diesel boiling range fractions) are liquids. Large components like asphalt and wax are solids (Perry et al., 1984). Efficient processing of all three phases at STP in large volumes, as required in a refinery, is not feasible. Because of the intrinsic molecular properties in crude oils (and/or refined products), processing them at STP is uncertain, irrespective of the energy source (radiation or conventional thermal processing) used to refine them.

Brief Description of Thermal Crude Oil Processing (Conventional Refining)

Refining crude oil principally involves separation of molecules by boiling point, along with some breaking and making of bonds (molecular rearrangement - chemical reactions). In order to begin the refining process, crude oil is first water washed to remove water soluble components. The washed crude is then sent to a furnace for preheating. The extent of energy expended in the preheating step depends on the crude viscosity. More viscous crudes require more heat in order to flow. The preheated crude is then sent to the distillation tower, where the energy expended to distill it depends on the compounds contained in the oil. Significant energy input for heat to conduct a distillation, cooling, compression, etc is required for refining to occur. Chemical reactions can be exothermic (giving off heat), but the vast majority are endothermic (requiring heat). Heat management and energy conservation are optimized to maximize product yield, meet product specifications and produce the highest economic return. Within narrow limits, refineries can alter processing configurations to adjust for changing feedstocks and product demand, while still meeting product specifications. Figure 1 shows a very simplified schematic of a refinery streams.

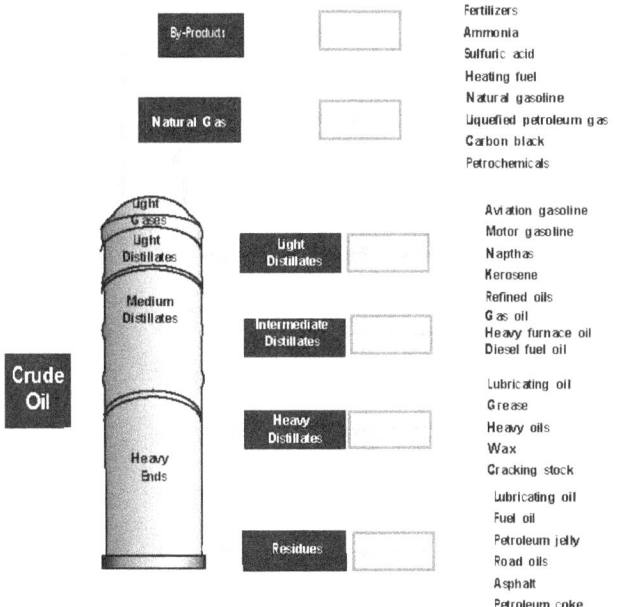

Figure 1: Simplified Schematic of Refinery Showing Principal Products

3

Figure 2 shows a simplified refinery with temperature cut ranges (boiling point ranges) for product streams from the crude distillation unit and vacuum distillation tower, all temperatures and pressures well above STP. A more inclusive description of refinery operation and units are available in the June 2000 National Petroleum Council (NPC), report to the Secretary of Energy entitled "U. S. Petroleum Refining, Assuring the Adequacy and Affordability of Cleaner Fuels." Reviews of conventional refining are available as part of the U.S. Department of Labor, Occupational Safety & Health Administration's (OSHA) technical manual on petroleum refining processes at: www.osha.gov/dts/osta/otm/otm_iv/otm_iv_2.html. DOE's Energy Information Administration (EIA) has an online refining tutorial and links to refining data at www.eia.doe.gov/pub/oil_gas/petroleum/analysis_publications/oil_market_basics/Refining. A unit by unit refinery analysis with chemistry, unit description and energy use is also available from the U.S. DOE Office of Industrial Technologies, Petroleum Industry of the Future, Energy and Environmental Profile of the U.S. Petroleum Refining Industry, (December 1998).

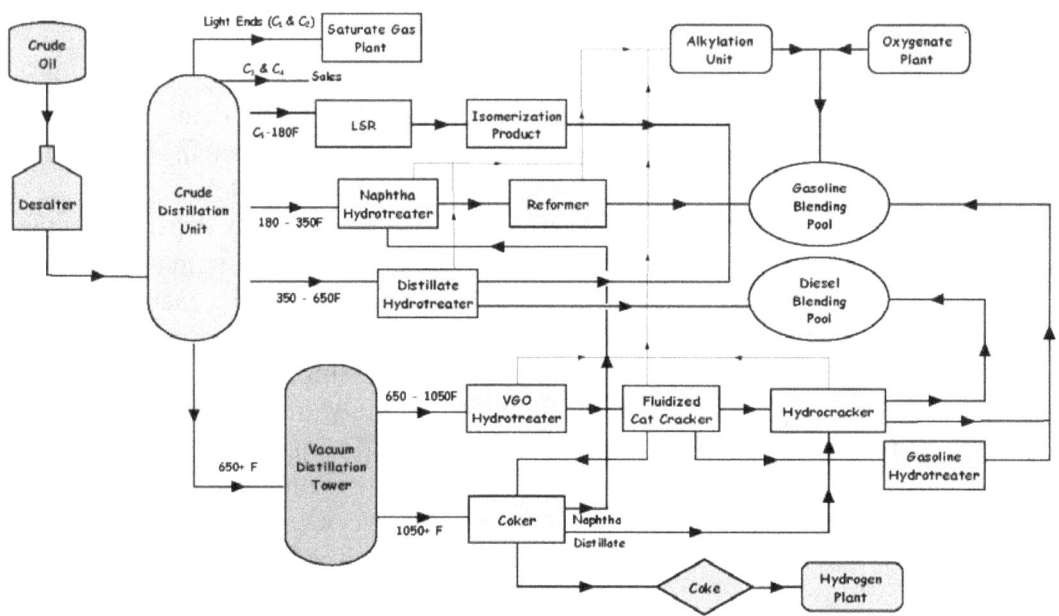

Figure 2: Simplified Schematic of Refinery Showing Principal Units and Temperature Cuts (Boiling Point Ranges) for Some Product Streams. (NETL)

For "cold cracking" it would be anticipated that refinery streams currently sent to a fluidized catalytic cracker (FCC) unit would be treated and not the entire crude oil stream. Chapters on FCC design, operation and economics can be found in many refining texts, including those specifically on FCC such as the FCC Handbook (2nd Edition) by Sadeghbeige (2000).

Radiation Processing of Crude Oil (Radiation-Thermal Conversion or Hydrocarbon Enhancement Electron-Beam Technology)

With irradiation of mixtures of hydrocarbons or crude oil, the changes in composition are reflected in gross changes to the sample viscosity, density, evolution of gases or formation of precipitates that can subsequently yield individual distillation cuts (gasoline, diesel, petroleum coke, etc.). There was no uniform temperature cut points. Characterization of changes in irradiated samples, which is not a standardized procedure, is usually evaluated by gas chromatographic analysis. Thus, higher severity radiation (time of exposure, intensity of

4

radiation, energy of the radiation, temperature of the sample being irradiated, pressure on the sample), as well as the energy source (neutron, positron, electron, Gamma-ray, X-ray, etc.) all contribute to changing reaction conditions. An attempt to correlate this set of variables to make performance predictions has not been successful. There are too many variables to simultaneously describe the limited series of published data. No consistent set of data was found in the open literature. A few literature references made comparisons of irradiated samples with sample subjected to conventional refinery processed samples. Many literature studies only compare the yields at various cut temperatures of irradiated crude samples with that of the untreated samples.

Review of the "four-page white paper sent to U.S. DOE" entitled "Cold Cracking of Petroleum Feedstocks" (Brainerd and Chappas, 2005) follows the trend in comparing an irradiated sample with an initial crude oil "feedstock". The conversion of predominately heavy crude to lighter products is evident from the results of the white paper. The cut points do not correspond with refiner's typical refinery cut ranges for gasoline, diesel, etc. No comments are made in the report on product stability. Discussion of the economics, in the Brainerd and Chappas paper, shows lower energy cost and consumption due to lower operating temperature. However, the paper gives no indication of the impact that lower temperature has on product yield or distribution. Extrapolating the curve to $0^{o}C$ (as in standard temperature of STP) shows the process will cost zero dollars, which is uncertain. The paper claims that the quality of the processed material is excellent, that processing efficiencies are high, and that the process is scaleable to industrial production rates. The claim that the process is scaleable is based on the fact that the product characterization method used is the same as that used by industry, according to Chappas (2005). The white paper continues with "This process, because of its low-temperature low-pressure operating environment, it can eliminate or reduce thermal processes such as coking and polymerization. By operating at ambient temperatures, the cold cracking process eliminates the energy required for preheating the feedstock." The white paper also claims that the technology's simplicity, safety, flexibility with regard to feedstock, high production rates, lower capital investment, operating costs, and the fact that the cold cracking process becomes inherently more economical than existing methods for oil processing. Evidence to support these claims is not obvious, either within the white paper itself or collectively within the literature reviewed.

Recently, advocates (including that of Brainerd and Chappas, 2005) have suggested use of radiation for petroleum processing as being less costly (capital and operating cost) and that an oil refinery could be constructed around the simplified schematic shown in Figure 3. An introduction to the basics of accelerators can be found at. www.lbl.gov/abc/wallchart/teachersguide/title_page entitled "Contemporary Physics Education Project, Nuclear Science — A Guide to the Nuclear Science."

Figure 3: Schematic of Hydrocarbon Enhancement Electron-Beam Technology (HEET) Refinery (Mirkin, Zaykina and Zaykin 2003)

The HEET refinery schematic is overly simplified. Significant additional processing units would be required to produce the slate of stable products listed in Figure 1.

Figure 3 shows a linear electron accelerator, an Electron-Positron Accelerator (LINAC) which is normally used for basic research projects and defense-related programmatic research. Most studies using LINAC's are experiments in fundamental nuclear, atomic, solid-state, plasma and particle physics or chemistry. A few industrial linear accelerators are used for initiating polymerization reactions or sterilization (killing microorganisms, fungi or spores). This is relatively safe because when the electron beam is switched off, there is no radiation in either the installation components or in the irradiated materials. (Tolsutun, Kuznetsov, Ivanov, Ovchinnikov and Svinjin) Worldwide, there are only a few companies that manufacture large electron beam systems, and each market a range of products with various energy delivery levels.

The literature contains many references on the use of radiation sources (forms) other than high energy electrons from a linear electron accelerator, including: neutrons, positrons, X-rays, Gamma-rays, beta-particles, and radioactive materials (e.g., fission products or daughter products from radioactive decay). Most can supply energies that are many orders of magnitude in excess of that required to break bonds in hydrocarbon molecules. Lower focused energies are also used for medical purposes (such as X-ray) or for determination of crystalline structures or molecular composition. Many patents and papers describe the use of "radiation" for thermal heating, nuclear process heat, or production of hydrogen.

Production of hydrogen from nuclear reactors for use in a refinery is an active field of research and development, both within U.S. DOE's National Laboratories and industry, but was not considered in this review. Likewise, for this review, use of excess heat from a nuclear reactor - thermal heating of petroleum crude oils was not considered. However, there are some U.S. patents (1950 to 1960's) on the use of crude oil and refined products as coolants (rather than water) circulated around a nuclear reactor. Early in this time period, no mention of potential problems of coking or deposition of high molecular weight materials on reactor walls was considered. However, later patents highlighted the problems of hydrocarbon degradation and focused on evaluation of highly aromatic refined products as heat exchange fluids - because under intense radiation, they are somewhat self healing and degrade much slower than paraffins or olefins. In the 1970's, significant work was conducted on the use of organically-cooled nuclear reactors and which aromatic or polycyclic aromatic molecules worked best as heat exchange fluids around reactors. Since polycyclic aromatics tend to deposit some coke, the focus of reactor coolants changed to the use of molten metals. Today, after years of research, water remains the most widely used fluid. With "cold cracking" the problem of radiation cracking is the reverse—you want to break molecules whereas in heat transfer, stable molecules are needed in the radiation field.

The National Nuclear Data Center (NNDC) at U.S. DOE's Brookhaven National Laboratory collects, evaluates, and disseminates nuclear physics data for basic nuclear research and for applied nuclear technologies. Part of the data is accessible online at www.nndc.bnl.gov. NNDC is a worldwide resource for nuclear data and specializes in the following areas that are pertinent to studies of radiation to process petroleum crude oils, including:

- Nuclear structure and low-energy nuclear reactions,
- Nuclear databases and information technology, and
- Nuclear data compilation and evaluation.

The radiation energy from a given source varies widely. Its application is like killing a mosquito, wherein one can use a fly swatter, baseball bat or semi-truck. They all get the job done, but some are overkill. The energies available from radiation are orders of magnitude higher than the activation energy needed for breaking a carbon-carbon, carbon-hydrogen, carbon-sulfur, or carbon-nitrogen bond. However, with conventional thermal crude oil processing, just enough energy to overcome the energy of activation is supplied to break the bond or conduct the chemical reaction.

Most nuclear data previously found only in specialized handbooks on nuclear data have become more easily accessible via the internet. Today, the U.S. DOE, through the Radiation Chemistry Data Center at the Notre Dame Radiation Laboratory (www.rcdc.nd.edu), is an information resource dedicated to the collection, evaluation, and dissemination of data characterizing the reactions of transient intermediates produced by radiation, chemical and photochemical methods. This data collection on individual reactions and radiation source energies shows that there are many radiation sources that can provide much more energy (many orders of magnitude) than necessary for most chemical bond breaking reactions.

Thermodynamic data on a large number of individual hydrocarbons, as well as classes of hydrocarbons, organo-sulfur and organo-nitrogen compounds, based on research conducted at DOE's National Institute for Energy and Research are included in the Chemical Property Databank in the Fifth edition of the Chemical Engineers Bible – The Properties of Gases and Liquids (Polingm, Prausnitz and O'Connell, 2002). Some thermodynamic data and chemical bond data have become available online at the National Institute for Standards and Technology at www.nist.gov, in addition to being published in the Journal of Chemical Engineering Data, the Journal of Physical Chemistry, and the Journal of the American Chemical Society or the Journal of Chemical Thermodynamics.

Literature Review

In public domain literature, the basic physics and chemistry of use of high radiation flux to crack hydrocarbon molecules is well known. This is based on studies of representative large molecular weight hydrocarbons - clean systems such as a series of linear alkanes, alkenes, substituted aromatic molecules, etc. However, literature on the application of radiation to crude oil, colloid or mixed systems is rare, except for polymer chemistry.

There are some "classic" literature references that summarize much of the work, including significant work on "Radiolysis of Liquid Hydrocarbons" conducted in Russia (Brodsky, et al, 1961; Földiák, 1980); "Aspects of Hydrocarbon Radiolysis" (Gäuman and Hoigne, 1968), "Radiation Chemistry of Hydrocarbons" (Földiák 1981), and "Radiolysis of Hydrocarbons" (Topchiev, et. al. 1964).

Földiák's "Radiolysis of Liquid Hydrocarbons" (1980) is a concise, illustrative paper on radiation chemistry of hydrocarbons and has extensive references. This is a good starting paper. It covers decades of research by hundreds of international researchers on hydrocarbon systems reacting to intense radiation to determine the mechanisms and reactivity trends of molecules. The paper principally covers irradiation of liquid samples. It covers product yields from the reactions of straight, branched and cyclic alkanes (paraffins), linear, branched and cyclic alkenes (olefins), and aromatic and substituted ring compounds.

The radiation chemical reactions of hydrocarbons are controlled by the structure and energy relations of the irradiated compound (pure hydrocarbon compounds, a series of pure compounds, or crude oil). Although there are differences between the energy necessary for given chemical reactions (breaking a chemical bond: C-C, C-H, C-S, C-N, or C-Metal atom) the energy required is about six orders of magnitude smaller than the energy levels of available radiation. The major difficulty is having a huge excess of energy at a given point in a sample of hydrocarbons. This leads to localized high concentrations of free radicals causing bond breaking, (cracking) polymerization and isomerization of molecules. Many of these reactions increase molecular weight and viscosity.

Radiation effect on organic molecules is really sensitive to molecule type. Aromatic compounds are known for high radiation resistance and can absorb excess energy of a considerable part of radiation-generated radicals. The Canadians looked at organic-cooled nuclear reactors (low pressure, low corrosion rates) and discovered that some of the multi-ring molecules were very stable or to be more precise, they tended to self heal when irradiated with neutrons. This suggests that one needs to consider crude oil irradiation with hydrogen overpressure or other hydrogen donors (radical quench) so chemical bonds that are broken do not just rejoin on themselves— polymerize. This is not thermodynamics — it's a kinetics problem.

Gamma-Rays and Radioactive Isotopes as Radiation Sources

Glockler (1947) presented a paper on "Controlled –Electron Reactions" as part of the Radiation Chemistry and Photochemistry Symposium at the University of Notre Dame in June 1947. His paper as well as numerous other papers presented some of the first insight of reaction mechanisms where high-energy particles are the initiators of a chemical reaction sequence. Hirschfelder (1947) at the same Symposium, covered some of the energetics of ionization caused radioactive particles and electrons interacting with organic molecules. Breger (1947) summarized five years of work at the Massachusetts Institute of Technology (MIT) on effects of alpha particles and deuterons on organic compounds. The work was one phase of a project sponsored by the American Petroleum Institute. Much of the early work focused on the role of radioactivity in petroleum genesis but also considered the reactions of radiation with various organic acids found in crude oil samples collected throughout the world.

Burrous and Bolt (1963) published their work on the use of refinery cuts as nuclear reactor coolants to see if they would "plate out" on heat transfer surfaces. The objective was to find a less expensive but stable heat transfer fluid. Their work analyzed use of gamma-ray irradiation of refinery cuts at $315\,^{\circ}C$ using a dosage of 5×10^9 rad. They compared the solubility and properties of the irradiated streams with radiolyzed terphenyls. The best of the refinery stream residues were less soluble and more intractable than the radiolytic polymer from terphenyls. Irradiation of refinery streams produced residue at faster rates than those from terphenyls. Thus, refinery stocks were at least as undesirable as those of polyphenyls. Since the lower cost of using a refinery stream as coolant does not overcome the effects of their decomposition, their use was not justified.

Skripchenko et al. (1986) published their experimental results on the decomposition processes that occur when petroleum fractions (boiling higher than $260^{\circ}C$) are irradiated with gamma-rays. The decomposition continues during distillation after previous exposure to gamma-rays. The fractions were sensitive to exposure to air (oxygen) which leads to condensation and production of products with increased molecular weight and reduced solubility in lighter hydrocarbons. Gamma-ray irradiation of a mixture of coal and the same petroleum fraction (at $300\text{-}400^{\circ}C$) yielded increased oil yield in the mixture, an increase in solubility in benzene and a reduction of oxygen containing groups in both the liquid and solid phases.

U.S. Patent 3,002, 911 by Sutherland and Allen (1961) teaches use of solid substrates for the transfer of energy from energetic–penetrating radiation to the adsorbed hydrocarbon to enhance conversion of the hydrocarbons compounds to lower molecular weight straight-chain and branched-chain structures. Sutherland and Allen (1961) reviewed the use of gamma-rays to convert olefins to polymerized material, and of converting paraffin (alkanes) to dimmers, olefins and gaseous products including CH_4 and H_2. They also show that the yield of various chemicals produced by irradiation of the organic compounds is different when a solid substrate is used versus when the same organics are irradiated in liquid form.

Cornwell (1993) patented a method of making and activating a copper-manganese (hopcalite) catalyst and activating the same (making more reactive sites on the catalyst) by irradiating the catalyst. Limited data is shown on performance of irradiated catalyst as compared to other commercial catalytic cracking catalysts. At temperatures below $80^{\circ}C$, the catalysts were inactive, and hydrocarbons tested were all light gases. Heavier hydrocarbons were not evaluated. They would not be anticipated to degrade even in the activated catalyst.

Coekelberg, et al. (1957) discussed methods of using radioactive solid waste from the nuclear industry as beta or gamma energy sources, including use of fission materials (e.g. U-235). The energy given off by these radioactive nuclides provides the energy sources for chemical reactions. Fission of an atom of U-235 has the energy of 162 MeV. The use of kinetic energy from heavy fission fragments (either dissolved in or dispersed in the hydrocarbon sample) depends on transfer of energy to the hydrocarbon. Coekelberg, et al. (1958) incorporated finely-divided, naturally-occurring uranium or uranium bonded to microporous solids in a neutron flux for achieving chemical reactions.

U.S. Patents 3,228,849 and 3,228,850 by Fellows (1966) describe the use of nuclear fission products to transfer energy for conducting chemical reactions, either in the presence or absence of porous or catalytic materials in crude oil dispersion. The patents describe a number of uses and references to the use of fission products as energy sources for hydrocarbon conversion.

The literature contains a large number of patents on the use of radioactive materials as catalyst support substrates or catalyst substrates for chemical reactions, including those that could occur in petroleum crude processing. Since oil refining requires high flow rates, there may be concern where the catalyst or substrate to support the catalyst are radioactive and where there is potential for carryover into the downstream separation equipment or where traces could be incorporated into the consumer product. Although there were a number of early patents and papers, use of the technology in today's more environmentally conscientious society is uncertain, and obtaining permits for such operation may be challenging.

Neutrons as Radiation Sources

U.S. Patent 2,905,606 by Long et al. (1959) describes conversion of hydrocarbons, including high boiling vacuum residuum, under high-pressure hydrogen and hydrogenation catalysts (platinum on alumina) in a neutron flux at temperatures from 50° to $700^{\circ}F$. The resulting hydrocarbons have increased hydrogen content in the lower boiling gasoline and diesel fractions.

U.S. Patent 2,905,607 by Long et. al. (1966), also describes conversion of distilled hydrocarbons in the presence of catalytic cracking catalysts (such as alumina) while exposed to a neutron flux. The resulting gasoline fraction has high isoparaffin content (high octane) and also makes high quality diesel fuel. The patent contains yields and refiner analysis for each fraction for a variety of crude oils.

U.S. Patent 2,905,608 by Long, et. al. (1966) describes conversion of distilled hydrocarbons in the presence of catalytic cracking catalysts such as alumna while exposed to a high gamma-ray flux. The resulting gasoline fraction has high isoparaffin content (high octane) and also makes high quality diesel fuel. The patent contains yields and refiner analysis for each fraction for a variety of crudes. The patent contains comparison of data with non-irradiated samples.

The above series of Esso patents contains some of the few comparison data sets for non-irradiated samples processed using conventional thermal refinery technology (of the time) and samples that were irradiated.

Pokonova and Meieshkov (1979) published their research on gamma-ray oxidation of asphaltenes to make phenolic compounds (used for oxidation inhibitors in mineral oils). Yields are poor compared to other routes to phenolic compounds.

Petukhov, et al. (1994) irradiated a black oil-tar mixture with electrons at $410^{\circ}C$ and 1 atmosphere and produced a liquid. They proposed using a neutron accelerator (not possible) as an irradiation source to process heavy hydrocarbons as part of a pilot-scale test.

Using thermal heat from nuclear reactors (normally used for generation of steam for electrical generation) or using a nuclear reactor to generate hydrogen from water may be more achievable than cold cracking of petroleum crudes by radiation. Neutrons would be anticipated to give a very different answer to "cold cracking - application of radiation to petroleum" because there is considerable energy loss via neutron-nucleus collisions, and thus one might well expect more hydrogen bond breakage.

As this literature review progressed, the focus shifted from all radiation sources (through a series of neutron sources, microwave and plasma generators, and gamma-rays and then to the use of high energy electrons) to the energy source proposed in the white paper by Brainerd and Chappas (2005), that initiated this literature search. Large hydrocarbon molecules typically found in crude oils are predominately composed of atoms of carbon and hydrogen, with lesser numbers of sulfur, nitrogen and metal atoms. The neutron cross-sections of carbon atoms are small and thus require a large number of neutrons for a single cracking event to occur. Neutron efficiency requires the presence of atoms (targets, like metal atoms) which adsorb neutrons well. Heavy crudes do not have enough good target atoms – as carbon, hydrogen, nitrogen and sulfur atoms are poor neutron absorbers. (Neutron News, 1992) However, electrons are much smaller and can be accelerated to much higher energies and higher flux to easily interact with hydrocarbon molecules.

High Energy Electrons as Radiation Sources

Reyes Lujan (1968) published his work on radiolysis of lubricating oil feedstocks, finished lubricating oils and high boiling point of petroleum residuum. The experiments used gamma-ray irradiation from Co-60 and 19 MeV electrons at 25 to 60°C and a dose range of 1 to 80 megarads. Garands (G) values for H and CH_4 and their variation with dose rate ranged between 0.00039 and 25 megarads. The irradiation study over a wide range of intensity showed some polymerization of the paraffinic lube oil feedstocks, whereas aromatics were much less affected by this level of radiation.

Mustafaev and Guileva, (1995) published "The Principles of Radiation-Chemical Technology of Refining Petroleum Residues" wherein they analyzed the scientific basis of radiation-induced thermal refining of heavy petroleum fractions (boiling point >300°C) using ionizing radiation, especially high energy electron and gamma-rays. Under these conditions (processing temperature of 430°C), approximately 71% of the heavy fraction was converted to naphtha and diesel fractions. The combined radiation and thermal treatment was calculated to be >10% more economic than conventional catalytic refining of the heavy fractions.

Aksenova, et. al. (1995) published a short summary of "Investigations on Radiation Processing in Kazakhstan" in which they looked at both gamma rays and electron-beams as energy sources for treating raw materials including crude oil. The paper is not quantitative or very descriptive. The emphasis was on using the energy sources to process metals or polymerizes monomers into polymers rather than application to petroleum refining.

Lykhterova, et. al. (1998) published a review of research on heavy oil refining (heavy oil, bitumen, pitch and coke). Research included use of electron beams, as well as ozone addition to the crude oil sample. The sulfur content was reduced (removed via formation of water soluble sulfones) and hydrocarbon molecules are transformed to lighter products (paper seems to be non-quantitative in that ozone reactions are known reactants to remove sulfur via sulfones).

In the last decade, researchers in Kazakhstan, along with some international colleagues, pursued a series of experimental studies of high-energy electron irradiation of hydrocarbons mixtures and crude oils. This is part of their research on radiation-thermal cracking (RTC) being performed at the Science Research Institute of Experimental and Theoretical Physics in Almaty, Kazakhstan. Research sponsorship has been through an international effort. Kazakh researchers have analyzed a number of hydrocarbon feedstocks under different radiation-thermal processing conditions. In most of the batch experiments, an oil sample was irradiated by 2 to 5 MeV (million electron volts) electrons generated by an electron accelerator, using current densities from 1 to 6 $\mu A/cm^2$. Different levels of processing (treatment) were provided by varying the temperature of the hydrocarbon samples, the irradiation total dose (usually in the range of 1-4 kG) and the rates of exposure (1-4 kG/s). The radiation chemical reactions of hydrocarbons are controlled significantly by the structures and the energy relations of the irradiated compounds. The differences between the energies of the chemical reactions are at least five orders of magnitude smaller than the energy of the radiation.

Zaykina, Zaykin, Mamonova and Nadirov (2001) in their paper entitled "Radiation-Thermal Processing of High-Viscous Oil from Karazhanbas Field" (one of the more complete and informative papers in the series by these authors) describe the impact of irradiation of a high–viscosity aromatic crude (Karazhanbas field, Kazakhstan) with high energy electrons. Irradiation at 350-450°C of a viscous, highly aromatic, low paraffin (1.5%), high sulfur (2%), high metal content crude with high energy electrons (2 MeV) showed that much of the aromatic content remained after irradiation, but the paraffinic components reacted to produce an increase in light products with an increase in branched alkanes – thus higher octane in the gasoline fraction. The high polycyclic aromatic rings proved more radiation-resistant then mono-cyclic aromatics. Substituents on the aromatic rings that were longer than ethyl tend to dealkylate and form toluene, xylene and styrene. The condensation reactions give products with increasing molecular weight and higher aromatization following the order: alkanes, resins, asphaltenes and coke.

Zaykina, Zaykin, Mirkin and Nadirov (2002) in the paper entitled "Prospects for Irradiation Processing in the Petroleum Industry" discuss the use of high energy electrons for processing of crude oils - petroleum refining. The G-values of 7000 (G-values for undefined compounds) seem high because normal G-values are 1-2 for many chemical reactions. Chain reactions can get to much bigger numbers, but 7000 seems high for molecule occurring in crude oils. Although comparisons are made with thermal cracking, the chemical distillation analysis, mass balance is not rigorous and therefore the comparison with what would be produced from conventional refining is not viable. No comments were made on changes in product quality when samples were subject to changing irradiation intensity.

Zaykina, Zaykin, Mirkin and Nadirov (2002) in the paper entitled "Radiation Methods for Demercaptanization and Desulfurization of Oil Products" uses ozonolysis to treat two crude oils, and then irradiated the samples at two different dosage levels. Quantitative analysis of the initial crudes, sample treatment, and results are not provided. General statements are made on utility, but they are not supported.

Zaykina, Zaykin, and Mirkin (2003) estimate the heat balances and energy consumption when using high energy electrons for hydrocarbon conversion. The paper does not adequately address energy consumption for hydrocarbon processing. The energy balance within the paper is not quantitative or supported. To compare heat with beam energy is quite specious. There are calculations of the efficiency of the conversion fossil fuels to electricity, transmission losses in the electricity, efficiency of the accelerator, energy loss in Bremsstrahlung gamma-rays being produced from the electrons, and relatively little of the typically refinery processing and it's efficiency is considered in converting crude oil to high quality refined products is included.

Mirkin, Aaykina and Zaykin (2003) published a non-quantitative, unsupported, general paper on hydrocarbon enhancement electron-beam technology (HEET). A variety of comparisons are made with conventional thermal refining of petroleum. The schematic of operation is overly simplified for refining, as it only focuses on a high energy electron beam focused on a flowing hydrocarbon stream. Figures seem comparable, despite the change of units. Chain lengths of 1000-2000 are not reasonable in crude oil systems for the G-values reported. Hydrocarbon chain lengths C-C are usually much less than 300 because the organic materials from which hydrocarbons are generated are rarely over 300 carbons long.

Zaykin, Zaykina and Nadirov (2004) in the paper entitled "Radiation-Initiated Cracking of Hydrocarbons and its Application for Deep Conversion of Oil Feedstock" describe the principal technological parameters that define energy consumption and economic efficiency of radiation-thermal cracking in crude oil processing. The paper is poorly supported and not rigorous.

In general, the samples of crude oils used by Kazakh researchers were common to Kazakhstan and the Caspian area. Crude oil assays (limited analysis compared to typical refinery assays of major international crude oils) showed that samples contained significant water (far more than the limit pipelines would accept) and it is not clear that the water was removed prior to the start of experiments. Some crudes were light, and separation of the water would have been simple; other heavier crudes had 10-15% emulsified water that would have required breaking the emulsion. Reactions of electrons with the high concentration of water molecules may provide a significant source of hydrogen, thus capping free radicals generated when the electrons impacted the organic constituents. Presence of water in the samples may have influenced the results (yields, product slate, viscosity reductions, etc.).

When high energy electrons were used to irradiate highly paraffinic crude oil from western Kazakhstan Kumkol field, polymerization and isomerization were observed (Zaykin, Aaykina and Silverman, 2004). Kumkol crude a high content of high molecular weight paraffins. Depending on the treating conditions, various product ratios were obtained. In general, they observed what would have predicted from earlier electron irradiation of mixtures of light paraffins studied by (Topchiev and Polak, 1962; Lavovsky, 1976). Irradiation of the paraffinic Kumkol crude showed high paraffin oils tend to polymerize. The experiments conducted on Kumkol oil samples had 10 to 15% of the mass as emulsified water, which may have significantly impact the results.

The Zaykin and Zaykina (2004) paper "Bitumen Radiation Processing" analyzed high energy electrons irradiating bitumen samples (from Shilikty and Mortuk oil fields of western Kazakhstan) with and without ozonolysis at 300°C. A comparison is made to thermocatalytic cracking yields of these hydrocarbon samples. The synthetic oil that is produced has a gasoline fraction with higher octane due to higher isoparaffin content. Quantitative mass balance to account for the other hydrocarbon fractions is lacking. Stability of the product gasoline fraction or other fractions is not reported.

The Zaykin and Zaykina (2004) paper entitled "Stimulation of Radiation-Thermal Cracking of Oil Products by Reactive Ozone-Containing Mixtures" describes the effect of flowing ozone and air through an oil sample, while simultaneously irradiating it with high energy electrons or gamma-rays in an attempt to reduce the required temperature for reaction and improve fuel quality of the light fractions. The gamma irradiation is the result of Bremsstrahlung gamma-rays from the 2MeV electrons. These reactions were conducted at room temperature. Higher yields of light fractions were obtained. However, stability of the ozonized product fractions is not addressed. No comparison is made with conventional thermal processing of the same crude oil sample.

The Zaykina, Zaykin, Yagudin and Fuhruddinov's (2004) paper entitled "Specific Approaches to Radiation Processing of High-Sulfuric Oil" describes treating high sulfur (sulfur content higher than 3% by weight) oils, which are usually very heavy, with ozone at room temperature and subsequent irradiating them to desulfurize the light oil fractions. Typically, heavy crudes contain sulfur in a variety of molecules with dibenzothiophene and substituted dibenziothiophene molecules occurring in the diesel boiling range. Thiophene and lighter sulfur containing molecules occur in the gasoline boiling range. The approach of ozonizing the sample and making some sulfoxide and sulfone molecules is well known, with thiophene more easily oxidized than benzothiophenes and much easier than dibenzothiophenes or substituted dibenzothiophenes. The lower molecular weight sulfones and sulfoxides can be removed by water washing of the sample. When water washed and distilled, the low boiling fractions traditionally have very little sulfur, and the higher boiling fractions contain significant sulfur as they are not reacted out of the mixture. However, these researchers chose to irradiate the entire sample after ozonolysis, and then distill the fractions. This resulted in the low boiling fractions showing little sulfur, and the high boiling fractions showing increased sulfur. No comparisons are made with conventional water wash treatment for removal of sulfones to determine if irradiation in fact had made any significant impact. Since electron irradiation does not convert sulfur to other atoms, the total sulfur remains the same, just redistributed in different boiling fractions. Accounting for all the sulfur in each fraction is needed. No product stability analyses were reported on the fractions.

Comments on Recent High Energy Electron as Radiation Source Papers

Within most of the recent Kazakh papers, there is a paucity of experimental details, so judgment about the experimental quality of the results cannot be made. Nearly all the references to processing of crude oils with radiation failed to address the product stability issue. Recombination of radicals, formed when bonds break, could allow recombination of the molecules into higher molecular weight molecules. Reactions continue long after the sample has been treated. A radical quench such as treatment with a high concentration of hydrogen would be needed to mitigate continuing polymerization. Without a quench, the higher molecular weight molecules – solids or resins, fall out on standing and plug or gum fuel systems. The reduction in sulfur reported using high energy electrons should be nearly the same as that obtained using thermal distillation in a conventional refinery. The data reflects sulfur distribution based on the structure of the compounds, with gasoline factions having the lowest and heavy bottoms having the highest concentration of difficult to remove sulfur species.

Interviews

All the interviewed refining researchers, refinery operators and refining planning personnel were very skeptical of using radiation for processing crude oils. They all believed that operation a refinery at STP is uncertain. They expressed concern that incorporating a "nuclear or radiation" unit into a refinery and that the acceptance of such idea by the public, their management or investors is uncertain. They felt than any increments resulting from this technology would be difficult to implement in an existing refinery and that use of radiation could be risky. Refiners

14

are very reluctant to be the "first or even the first few" to implement a technology that has not been commercially proven at a number of locations. Technology implementation into a refinery carries a high risk in that permitting, construction, and startup require many years, integration with the rest of the refinery and hundreds of millions to billions of dollars of investment. There are dozens of examples of very efficient technologies (producing higher yield at lower cost) that have years of pilot plant applications and well documented experience that have never made it to commercial scale. This is because the perceived risk of being first is too high. Potential fines from producing products that are off specifications (resulting in requirement for reprocessing or fines from environmental regulators), less than anticipated product yield, higher processing cost than anticipated, loss of market for product, time delays to correct problems, or the problem of building a unit that becomes a stranded liability makes refiners more adverse to risk - cautious investors. It is the impression of most of the refiners interviewed that radiation processing of crude oil would exceed that threshold.

A number of nuclear engineers, scientists, and educators, who are familiar with some of the work that has been conducted on radiation of organic molecules, were also interviewed. Discussions with them revealed that some studies were conducted in the 1950-1960's, both in the U.S. and internationally. Some of them were aware of the research programs in Russia and the more recent studies conducted by their scientists. A few nuclear scientists offered comments on a number of recent papers published on the use of electron beams for processing organic molecules, including crude oil. Their general comments were that the recent papers have sparse details and the research studies in recent years have not been as deep as those of the 1950-60-70's.

Discussion

Use of radiation for processing materials has been evolving since the introduction of the technology nearly fifty years ago. Crosslinking plastic materials, sterilizing medical products and preserving foods were some of the earliest applications. The application to a broader area has been slow. Many of the papers describing these applications have been published in the proceedings of the thirteen International Meetings on Radiation Processing (Radiation Physics and Chemistry, 1977- 2005).

Permitting a new nuclear plant and a new petroleum refinery in the U.S is a lengthy and expensive process. The last new grassroots refinery in the U.S was constructed in 1976 and the last new grass roots U.S. nuclear plant finished construction and started power production in 1996, after 24 years of permitting, planning and construction.

Neutrons as Radiation Source

Use of neutrons in the application of radiation to petroleum refining today is uncertain for a number of reasons:
- Nuclear reactors (neutrons) are very expensive.
- Neutrons tend to activate certain materials and make them radioactive. While low levels of radioactivity might have been considered acceptable in the 1950s, its acceptability today is uncertain.

Japan's experience in permitting nuclear reactors is much shorter (< 10 years) as compared to the U.S. Recent experience outside the U.S. shows that once permits are obtained it can be as short as four years from pouring first concrete to delivery of electrical power.

Radioactive Sources

Gamma-ray sources, such as Co-60, require extensive shielding. Radiation is always being emitted, and exposure is controlled by opening and closing the window on the containment vessel. Co-60 or any number of long-life radioactive sources emitting neutrons, Gamma-rays, X-rays, Beta particles, etc., some with half-life of thousands of years or hundreds of thousands of years, can provide a constant radiation source for very long periods, much longer than the containment vessel. Therein is one of the major problems. Once constructed, the radiation is usually low cost over a few decades or centuries, and will outlast the other processing components or the containment vessel. Over the life of the source (thousands of years) shielding degrades, and this occurs before the radiation level reaches acceptable levels for disposal.

Linear Electron Accelerators as a Radiation Source

One advantage of linear accelerators is that when the electrical energy is turned off, there is no residual radiation. During the last 20 years, technology advancements and lower operating costs have given big high-efficiency electron beam guns (linear accelerators) some advantages. One can dial in the energy of the electron and chose an energy that maximizes gain for a specific chemical reaction in the target molecule. The next generation of particle accelerators may not require the large space and cost less than the billion dollar units that exist today. However, laser electron accelerators may offer a cheaper and smaller alternative. Huge electric fields in laser-produced plasmas have accelerated beams of electrons close to the speed of light - an important step towards the development of a working laser electron accelerator, which can cost millions of dollars and occupy less than a 100 meters^2 of space. Both military and industrial application of electron beam technology has become closer to economic application. Whether this could be a commercial technology for cold cracking remains uncertain because of the broad mixture of molecules and possible primary and secondary reactions. Advances in electron beam technology make most comparisons with Co-60 Gamma-ray irradiation sources and the economics outdated (Morrison, 1989). Comparisons of radiation sources have to take into account effective penetration and energy levels. Electron beams used for food processing cannot penetrate food products more than 3 inches thick. In contrast, gamma-rays and x-rays can penetrate an entire pallet load of food products.

Batch vs. Continuous Flow Test of Radiation Processing of Crude Oil

A proposal (# K-930 (2004)) submitted to the International Science and Technology Center (ISTC, Russia), is for construction and continuous operation of a "large-scale experimental facility" for processing natural bitumen with high-energy electron beams. The proposed production rate of up to 200 kg per hour (1.46 barrels/hour, 35 bbl/day) for processing bitumen (API gravity of less than 10° and viscosity of greater than 100,000 cp – a semisolid at STP) by high-energy electron beams will attempt to "improve yields and quality of commodity oil products. This proposed project would permit Kazakh researchers to test, for the first time, their electron beam system with a flowing (continuous) hydrocarbon system. No details of the experiments to be performed are given in the proposal. Thus, details of yields, energy consumption, energy cost, product stability, product quality, or overall economics that would allow for comparison of conventional refining with radiation-thermal conversion are still needed. Even a smaller scale, continuous flow system with appropriate chemical analysis of samples could provide insight into the feasibility of radiation processing of crude oil. Significant refining/processing tests, chemical and economic analyses would be required to make viable comparisons of the two technologies (thermal or thermal-catalytic refining versus radiation processing) when using a well-defined crude oil. High aromatic content, viscous bitumen would not be the first choice of crude oil to test in a flowing system.

Permitting of Refinery and Radiation Generators

In the U.S., radiation sources are permitted by the U.S. Nuclear Regulatory Commission (NRC) and the Occupational Safety and Health Administration (OSHA). In recent years, the U.S. Environmental Protection Agency (EPA), the Department of Energy, and the Department of Homeland Security have been given additional oversight authority. NRC has the authority to regulate source, by-product, and certain special nuclear materials. OSHA's authority to regulate radiation sources covers all radiation sources not regulated by NRC, including X-ray equipment, accelerators, accelerator-produced materials, electron microscopes, betatrons, and some naturally-occurring radioactive materials. Both NRC and OSHA have jurisdiction over occupational safety and health at NRC-licensed facilities. Because it is not always practical to sharply identify boundaries between the nuclear and radiological safety issues, a coordinated inter-agency effort is used to ensure the protection of workers and the public.

As mentioned earlier, permitting a dedicated nuclear source in or near a refinery may be expensive and lengthy. Its public acceptance may be challenging, as the "not in my back yard" (NIMBY) attitude prevails for both the nuclear and petroleum industries. Use of a high energy electron accelerator may appeal more to a refinery, but its public acceptance may be more challenging. Significant energy savings, increase in yield, and increased public education and awareness nay be needed for cold cracking technology can compete with conventional refining.

Energy Consumption

There are a number of estimates of energy consumption for crude oil conversion via irradiation with high energy electrons. More recent tests used a dose of about 50 kGy (50 kJ/kg). This is equivalent to about 0.014 kW/kg of crude oil, which at $0.06/kW for electricity would yield a radiation exposure cost (processing cost) of about $0.0008/kg. (Walter Chappas, personal communication, November 25, 2005.) This implies that the processing cost of a heavy crude oil sample (10°API gravity, density of 1 kg/liter) via irradiation with high energy electrons would be approximately $0.13/barrel. This is significantly lower that conventional refining cost of about $4/barrel, assuming $30/barrel heavy crude oil price. However, it is difficult to compare the cost of radiation processing with that of conventional refining. The conventional refining cost given here include non-volume related expenses (wages and salary, benefits, equipment maintenance, etc) and volume related expenses (catalysts and additives, utilities, fuels, royalties), while the radiation processing cost is just the radiation energy cost. For a fair and accurate comparison, information on the other expenses similar to those given for conventional refining is necessary. Besides, the extent of conversion of the sample achieved when irradiated is not known. A pending patents by Chappas et. al. may provide some of the information when it is issued and additional details become public.

Other previous studies on energy consumption of cold cracking provided little or no information to back their claims. The energy consumption claims by Mirkin (2003) for construction and operation (40 to 65% less energy) are unsupported. Aksenova's (1995) energy statements are also unsupported. The Zaykina, Zaykin, and Mirkin (2003) paper on energy consumption, on which much of the economics is based, is not rigorous and does not include significant cost factors or energy conversions to converting crude oil to refined product or separating the refined product to commercial quality streams.

Capital and Operating Cost of a High Energy Electron Beam Source

Before addressing the capital and operating costs of large electron beam systems, a description of the technology is provided here. Discussions with IBA Industrial, one of the few manufacturer's of large electron beam systems, considered the use of their unit, called the Rhodotron TT1000, as the electron source for application in cold cracking of petroleum (Herer, 2006). The TT100 can provide about 100 mA of electrons in the 5 to 7 MeV (penetration) range (hence 500 to 700 kW). The accelerator was developed for high-powered X-rays (for high-volume, high-dose food irradiation), and is considerably more powerful than any other electron beam system available anywhere. It can also be used as an electron beam. When operating at full beam power, 700 kW, the electron beam system will directly consume about 1,400 kW, and may require another 25 to 30% in terms of kW for cooling. (This applies to the Rhodotron TT1000; smaller, lower energy electron beam sources will consumer more power per kW of energy output.)

A schematic for the Rhodotron electron beam accelerator is shown in Figure 4. Electrons are accelerated as they pass through a properly oriented electrical field. The electrical field in the single coaxial-shaped cavity of the Rhodotron is radial and oscillates at a frequency of either 107.5 or 215 MHz, depending on the Rhodotron model. During each of these radio-frequency (RF) cycles, electrons are fired by the electron gun and introduced into the cavity when the electrical field is such that it will accelerate the electron inwards towards the hollow coaxial cylinder in the center. The electrons then pass through openings in the inner cylinder, while the electrical field is reversing, and on emerging from the inner cylinder, the electrons are further accelerated towards the outer cavity wall under the influence of the now reversed field.

Photo Removed Due to Copyright Restrictions

Figure 4: Schematic of Rhodotron Electron-Beam System
(Courtesy of Ion Beam Applications, 2006)

Using deflection magnets, the electrons can be reintroduced into the main body of the accelerator for additional crossings of the cavity. To achieve an energy output of 10 MeV, the large-diameter Rhodotron models use 10 passes through the cavity, and smaller diameter Rhodotron models use 12 passes.

In terms of throughput, each kW of beam output can treat (at 70% capture efficiency) about 2500 kg/hr to a dose of 1 kGy (3600*0.7) (Herer, 2006). Hence, for a treatment dose of 15 kGy, using a 700 kW beam, one can treat 2500*700/15 = 16,666 kg /hour. The variables might differ, but the same equations would apply. For a 10° API crude (density of 1 Kg/l), this would translate to 3,355 bbl/day. If the electrical energy used for cooling the electron source can be replaced (heat exchanged) by using oil to cool the system (and thus preheating the oil) some of the 25% to 30% in terms of kW for cooling could be recovered. Thus, the 700 kW beam power consumes 1,400 kW for the beam plus possibly less than 420 kW for cooling the electron beam source (1,400 + 420 = < 1820 kW). It may also be possible to increase the 70% capture efficiency by configuration of the exposure cell through which the oil is being pumped when it is irradiated.

Total electron beam facility pricing (once shielding and conveyance is included) will not scale down proportionately, especially at lower powers. For small volume refineries, there might be some saving by lowering the MeV (penetration) of the system. The approximate price of such a TT1000 unit is $5,500,000, including accelerator and scan horn (installed price). The rest of an irradiation system (shielding, product conveyance, etc.) would need to be built around this, which may be more than double the cost for a system.

Operating costs of electron beam facilities of various sizes and beam penetrating power are available from the Ion Beam Applications website at http://www.iba-worldwide.com.

Product Stability

Refined products (gasoline, diesel, aviation fuel, etc.) are usually less stable (shorter storage life) than crude oil. In some applications, degradation of fuel may only plug fuel filters or foul fuel injectors. There are additives that can extend the working life of fuels. In more critical applications such as aviation, fuel quality and fuel stability are primary objectives. Many fuels continue to chemically react, even after being conventionally refined. Significant While the literature on radiation conversion of petroleum (crude oil or simple compounds) has numerous references product stability after treatment, most of the studies address short term stability effect. None of them is quantitative or examines long term stability effects. Many reports indicate that the product continues to change color or viscosity.

Conclusions

Available literature on application of radiation technology has been reviewed and analyzed. It is obvious from the literature review and private discussions with experts that application of radiation to petroleum research is not new. The review also indicates that processing crude oil in a cost-effective and environmentally safe manner using radiation is uncertain. Operating cold cracking at STP may pose some material handling problems associated with multi-phase (gas, liquid and solid) systems. From the literature data, the economics of "cold cracking" are uncertain. The few cold cracking "beaker tests" conducted were not comprehensively studied. There is too little quantitative information to make comparisons with conventional refining. It is therefore unclear as to whether there are energy and economic benefits when cold cracking is used.

References

Aksenova, T. I., D. K. Daukeev, B. M. Iskakov Yu. A. Zaykin, N. R. Mazhrenova and A. S. Nurkeeva. "Investigations on Radiation Processing in Kazakhastan." Radiation Physics and Chemistry, V. 46, Issue 4-6, pp. 1401-1404, 1995.

Brainerd, G. R and W. J. Chappas. "Cold-Cracking of Petroleum Feedstocks" an undated white paper submitted to U.S. DOE presumably in 2005.

Brainerd, G. R. Personal Communication, 2005.

Breger, I. A. "Transformation of Organic Substances by Alpha Particles and Deuterons". Published in the Proceedings of the Symposium on radiation Chemistry and Photochemistry, University of Notre Dame, Notre Dame, Indiana, June 24-27, 1947. Reprinted in J. Physical and Colloid Chemistry V. 52, 3, pp, 551-563, and 1948.

Brodsky, A. M., N. V. Zvonov, K. P. Lavrovsky and V. B. Titov. "Radiation-Thermal Conversion in Oil Fractions." Neftekhimiya (Oil Chemistry - Russian) V.1, N3, pp. 370-381, 1961.

Burrous, M.L. and R.O. Bolt. "Petroleum Refinery Streams as Nuclear Reactor Coolants-Radiolytic Product Investigations." Internal Atomic Energy Agency, Report No. TID-19440, 19p. 1963.

Chappas, W. J., Personal Communication, 2005.

Coekelberg "Some Future Aspects of Radiochemistry." Belgishe Chemische Industrie, V. 22 No. 2, pp. 153-164, 1957.

Coekelberg et. al., "Investigation of a Nuclear Fuel Making it Possible to Use the Kinetic Energy of Fission Products for Chemical Synthesis." V. 29 pp. 424-32, Proceedings of Second International Conference on the Peaceful Uses of Atomic Energy, 1958.

Cornwell, J. H. "Catalyst for Molecular Catalytic Cracking of Heavy Hydrocarbons at Ambient Temperatures, and Method of Making the Same." U.S. Patent 5,238, 897 assigned to North Carolina Center for Scientific Research, 1993.

Dougherty, D., Nerac, Inc. A Literature Review Report Prepared for Thomas G. Peterson, August 24, 2005.

Fellows, A. T. "Utilization of Nuclear Fission for Chemical Reactions." U.S. Patent 3,228, 849 assigned to Scony Mobil Oil Corporation, 1966.

Fellows, A. T. "Chemical Conversion in Presence of Nuclear Fission Fragments." U.S. Patent 3,228, 850 assigned to Scony Mobil Oil Corporation, 1966.

Földiák, Gábor. "Radiation Chemistry of Hydrocarbons." ISBN: 0444997466 Elsevier Scientific Publishing, 1981.

Földiák, Gábor. "Radiolysis of Liquid Hydrocarbons." Radiation Physics and Chemistry, V. 16 Issue 6 pp. 451-463, 1980.

Gäuman, T., and T. Hoigne. (Eds.) "Aspects of Hydrocarbon Radiolysis." Academic Press, London, 273 p. 1968.

Glockler, George. "Controlled – Electron Reactions. Published in the Proceedings of the Symposium on radiation Chemistry and Photochemistry, University of Notre Dame, Notre Dame, Indiana, June 24-27, 1947. Reprinted in J. Physical and Colloid Chemistry V. 52, 3, pp. 451-457, 1948.

Herer, Arnold. Cost and Operation of Operation of Electron Beam Systems, Rhodotron TT1000, IBA Industrial, Personal Communication, March 2006.

Hirschfelder, J. O. "Chemical Reactions Produced by Ionization Processes." Published in the Proceedings of the Symposium on radiation Chemistry and Photochemistry, University of Notre Dame, Notre Dame, Indiana, June 24-27, , 1947. Reprinted in J. Physical and Colloid Chemistry V. 52, 3, pp. 447-450, 1948.

Ion Beam Applications (IBA) website at http://www.iba-worldwide.com.

Lavovsky, K.P. "Catalytic, Thermal and Radiation Chemical Conversion in Hydrocarbons." Nauka, Moscow, pp. 255-263 and pp. 312-373, 1976.

Lavovsky, K.P., A. M. Brodsky N. V. Zonov and V. B. Titov. "Radiation-Thermal Conversion of Oil Fractions." Neftelhimiya (Oil Chemistry) V. 3, Issue 3, pp. 370-383, 1961.

Long R.B., H. J. Hibshman, J. P. Longwell and R. W. Houston. "Conversion of Hydrocarbons in the Presence of Neutron Radiation, Hydrogen and a Cracking Catalyst. " U.S. Patent 2,905,606 assigned to Esso Research and Engineering, 1959.

Long R.B., H. J. Hibshman, J. P. Longwell and R. W. Houston. "Conversion of Hydrocarbons in the Presence of Neutron Radiation and a Cracking Catalyst. " U.S. Patent 2,905,607 assigned to Esso Research and Engineering, 1959.

Long R.B., H. J. Hibshman, J. P. Longwell and R. W. Houston. "Conversion of Hydrocarbons in the Presence of Gamma-ray Radiation and a Cracking Catalyst." U.S. Patent 2,905,608 assigned to Esso Research and Engineering, 1959.

Lykhterova, N.M., V.V. Lunin, and M. V. Lomonosova. "Unconventional Methods for Processing of Heavy Petroleum Feedstocks." Netf. I Gaz. V. 6 pp. 3-5, 1998. (In Russian)

Mirkin, G., R. F. Zaykina and Yu. A. Zaykin. "Radiation Methods for Upgrading and Refining of Feedstock for Oil Chemistry. Radiation Physics and Chemistry, V. 67, Issues 3-4, pp. 311-314, 2003.

Morrison, R. M. "An Economic Analysis of Electron Accelerators and Cobalt-60 for Irradiating Food." U.S. Department of Agriculture. Technical Bulletin No. (TB-1762) 48 pp, June 1989.

Mustafaev, I. and N. Guileva "The Principles of Radiation-Chemical Technology of Refining Petroleum Residues" Radiation Physics and Chemistry, V. 46, Issue 4-6, pp. 1313-1316, 1994.

National Petroleum Council. Appendix A-D on Refining (used with permission) Originally published as the June 2000 National Petroleum Council report to the U.S. Secretary of Energy entitled "U. S. Petroleum Refining, Assuring the Adequacy and Affordability of Cleaner Fuels."

Neutron News, Vol. 3, No. 3, 1992, pp. 29-37 (Neutron cross section data.)

Perry, R. H., D. W. Green and J.O. Maloney. "Perry's Chemical Engineers Handbook." Sixth Edition, McGraw-Hill, 1984.

Peterson, T.G., Personal Communication, 2005.

Petukhov, V. K., A. I. Chekushin, and Yu. A. Zajhin. "Use of Radiation Technologies for Oil Processing." Ehkspress-Informatsiya. Novosti Kauki Kazakhstana, V. suppl. 1, pp. 23-25, 1994. (In Russian)

Poling, B.E., J. M. Prausnitz and J. P. O'Connell. "Fifth Edition of the Chemical Engineers Bible – The Properties of Gases and Liquids." McGraw-Hill, 2002.

Pokonova, Yu. V. and S. P. Meieshkov. "Preparation of Petroleum Asphatols." Zhurnal Prikladnol Khimii, V52 No. 10, pp. 236, 1979. (In Russian)

Proceedings of the Thirteen International Meetings on Radiation Processing, Radiation Physics and Chemistry, Pergamon Press, Elsevier Science, Inc., Tarrytown, New York. V9, 14, 18, 22, 25, 31, 35, 42, 46, 52, 57, etc. 1977-2005.

Reyes Lujan, Javier. "Radiolysis of Petroleum Products." Revista Mexican de Fisica V. 17, Issue 1, pp. 1-17, 1968. (In Spanish)

Sadeghbeigi, Reza. "Fluid Catalytic Cracking Handbook" 2nd Edition, Gulf Publishing, 2000.

Suib, S. L. and Z. Shang. "Low Power Density Plasma Excitation Microwave Energy Induced Chemical Reactions." U.S. Patent 5,131,993 assigned to The University of Connecticut, 1992.

Skripchenko, G.B., V.I. Sekrieru, N.K. Larina, Z.S. Smutkina, O.K. Miesserova and V.A. Rudol. "Effect of Irradiation on Heavy Petroleum and Coal Products." Kimiya Tverdogo Topliva. V.5, pp. 55-59, 1986. (In Russian)

Sutherland, J. W. and A. O. Allen. "Radiolysis of Organic Compounds in the Adsorbed State." U.S. Patent 3,002,911 assigned to the U.S. Atomic Energy Commission, 1961.

Tolstun, N. G., V. S. Kuznetsov, A. S. Ivanov, V. P. Ovchinnikov, and M. P. Svinjin. "The 3 MEV, 200 KW High Voltage Electron for Industrial Application." Publication of the D.V. Efremov Research Institute of Electrophysical Apparatus, 189631, St Petersburg, Russia.

Topchiev, A. V., L. S. Polak, and R. A .Holroyd. "Radiolysis of Hydrocarbons" 1964 English ed. New York, Elsevier. LCCN: 63-19826 Summary of Research Conducted in the Radiation

Chemistry Laboratory of the Institute for Petrochemical Synthesis of the Academy of Sciences of the USSR during the years 1957-1961."

Topchiev, A. V. and L. S. Polak. (Eds.) "Hydrocarbon Radiolysis." Academy of Science, 208p. Moscow, USSR, 1962.

Zaykin, Yu. A., R. F. Azykina, and G. Mirkin. "On Energetics of Hydrocarbon Chemical Reactions by Ionizing Irradiation." 10 'Tihany' Symposium on Radiation Chemistry, Sopron (Hungary), Program Abstracts, August 31- September 5, 2002, p.140.

Zaykin, Yu. A., R. F. Azykina, and G. Mirkin. "On Energetics of Hydrocarbon Chemical Reactions by Ionizing Irradiation. Radiation Physics and Chemistry, V. 67, September-October 2004, pp. 305-309, 2003.

Zaykin, Yu. A. and R. F. Zaykina. "Bitumen Radiation Processing." Radiation Physics and Chemistry, V. 71, Issues 1-2, September-October 2004, pp. 471-474, 2004.

Zaykin, Yu. A., R. F. Zaykina and N. K. Nadirov. "Radiation-Initiated Cracking of Hydrocarbons and its Application for Deep Conversion of Oil Feedstock." Oil and Gas (Kazakhstan), V. 4, Issues 24, pp. 47-57, 2004.

Zaykin, Yu. A., R. F. Zaykina and Joseph Silverman. "Radiation-Thermal Conversion of Paraffinic Oil." Radiation Physics and Chemistry, V. 69, Issue 3, pp. 229-238, 2004.

Zaykin, Yu. A., and R. F. Zaykina. "Bitumen Radiation Processing." Radiation Physics and Chemistry, V. 71, pp. 469-472, 2004.

Zaykin, Yu. A., and R. F. Zaykina. "Stimulation of Radiation-Thermal Cracking of Oil Products by Reactive Ozone-Containing Mixtures." Radiation Physics and Chemistry, V. 71, pp. 473-476, 2004.

Zaykina, R. F., Yu. A. Zaykin, G. Mirkin and N. K. Nadirov. "Prospects for Irradiation Processing in the Petroleum Industry." Radiation Physics and Chemistry, V. 63, Issues 3-6, pp. 617-620, 2002.

Zaykina, R. F., Yu. A. Zaykin, T. B. Mamonova and N. K. Nadirov. "Radiation-Thermal Processing of High-Viscous Oil from Karazhanbas Field." Radiation Physics and Chemistry, V. 60, Issue 3, pp. 211-221, 2001.

Zaykina, R. F., Yu. A. Zaykin, Sh. G. Yagudin and I. M. Fahruddinov. "Specific Approaches to Radiation Processing of High-Sulfuric Oil." Radiation Physics and Chemistry, V. 71, Issues 1-2, pp. 467-470, 2004.

Zaykina, R. F., Yu. A. Zaykin, Sh.G. Yagudin and I.M. Fuhruddinov. "Specific Approaches to Radiation Processing of High-Sulfuric Oil." Radiation Physics and Chemistry, V. 71, pp. 465-468, 2004.

www.ingramcontent.com/pod-product-compliance
Lightning Source LLC
Chambersburg PA
CBHW081317180526
45170CB00007B/2750